STRAIGHT LINES IN A CURVED UNIVERSE

Terrence Howard's Exploration of Space and Time

Oteren.Fredrick

DISCLAIMER

The substance introduced in this book, "Straight Lines in a Bended Universe: Terrence Howard's Investigation of Existence," mirrors the understandings and points of view of the writer and givers. The hypotheses and thoughts talked about, especially those proposed by Terrence Howard, address capricious perspectives that challenge laid out logical ideal models.

COPYRIGHT

TABLE OF CONTENT

INTRODUCTION

The universe, in its immense intricacy and magnificence, has long entranced researchers, scholars, and masterminds the same. How we might interpret the universe is outlined by key inquiries: What is the idea of reality? How do straight lines squeeze into a bended universe? These requests are hypothetical insights as well as are critical to appreciating the actual texture of the real world. In this investigation, Terrence Howard's commitments express an extraordinary and viewpoint inciting point of view, testing conventional thoughts and presenting creative thoughts regarding the construction of the universe.

Terrence Howard's Hypothetical Contributions

Terrence Howard, essentially referred to for his work as an entertainer, has likewise made critical commitments to logical idea, especially in the domain of hypothetical material science. His flighty speculations have started discussion and interest inside mainstream researchers. Howard's work stands apart in light of the fact that it combines imaginative knowledge with thorough logical request, overcoming any issues between unique ideas and viable comprehension.

At the core of Howard's hypotheses is the possibility that our traditional comprehension of reality might be essentially defective. His methodology challenges the customary Euclidean perspective on straight lines and level spaces, proposing rather a model where straight lines exist inside a bended structure. This point of view lines up with the more extensive logical comprehension of general relativity, which portrays what gravity means for the curve of room time. In any case, Howard expands this system with novel understandings that welcome further assessment and discussion.

The Convergence of Science and Reasoning in Howard's Work

Howard's hypotheses are not just logical yet additionally profoundly philosophical. He draws in with inquiries concerning the idea of the real world and our place inside it, which are fundamental to the two material science and power. By investigating the idea of occasion skylines and the ramifications of straight lines inside a bended universe, Howard's work meets with philosophical requests about the idea of presence and insight.

The idea of occasion skylines, for example, addresses the restrictions of our observational capacities and the limits of our insight. In astronomy, an occasion skyline addresses the limit encompassing a dark opening past which nothing can escape. Howard's investigation of this idea includes the actual ramifications as well as the philosophical consequences of such limits on how we might interpret the universe.

Reason and Extent of the Book

This book, "Straight Lines in a Bended Universe: Terrence Howard's Investigation of Room and Time," expects to give a top to bottom examination of Howard's commitments to how we might interpret the universe. It looks to enlighten how Howard's thoughts challenge and grow customary logical speculations, offering new viewpoints on space, time, and the basic construction of the real world.

The extent of this book envelops a few key regions:

1. Foundations of Room and Time: We will begin by looking at the fundamental ideas of occasion skylines and straight lines in bended spaces. This basic information is urgent for understanding the further developed speculations introduced by Howard.

2. Terrence Howard's Logical Journey: We will dig into Howard's experience, his persuasions, and the improvement of his hypotheses.

Understanding his scholarly excursion gives setting to his imaginative thoughts.

3. The Calculation of the Cosmos: This segment will investigate the numerical and hypothetical underpinnings of Howard's work. We will dissect how he deciphers the connection between straight lines and bended spaces, and how this connects with how we might interpret the universe.

4. The Nature of Occasion Horizons: We will examine Howard's perspectives on occasion skylines, contrasting them and laid out logical hypotheses and investigating their suggestions for how we might interpret existence.

5. Straight Lines in a Bended Universe: This part will zero in on Howard's remarkable point of view on straight lines inside a bended universe. We will look at his contentions, outlines, and the ramifications for how we might interpret all inclusive designs.

6. Exploring Existence Through Howard's Lens: Here, we will evaluate Howard's commitments to how we might interpret the space-time continuum and their effect on contemporary logical idea.

7. Criticisms and Controversies: No logical hypothesis is without its faultfinders. This part will address the discussions and discussions encompassing Howard's work, giving a fair perspective on the difficulties and reactions to his hypotheses.

8. Practical Applications and Future Directions: We will investigate the likely uses of Howard's speculations and think about future examination headings. This part will likewise think about Howard's vision for the eventual fate of reality studies.

9. Interviews and Insights: This section will remember discussions with Howard and experiences from specialists for the field,

offering a more private and itemized perspective on his hypotheses and their suggestions.

10. Conclusion: The last area will sum up Howard's effect on existence speculations, pondering the more extensive ramifications of his work and what's to come possibilities for additional investigation.

By drawing in with these points, this book expects to give a thorough outline of Terrence Howard's commitments to how we might interpret the universe. It will investigate how his inventive thoughts challenge and extend our ongoing logical and philosophical ideal models, offering new bits of knowledge into the idea of reality.

All in all, "Straight Lines in a Bended Universe" looks to overcome any barrier between ordinary logical speculations and creative points of view. It welcomes perusers to investigate the limits of our insight and think about additional opportunities for figuring out the universe.

Through a nitty gritty investigation of Howard's work, we desire to motivate further request and reflection on the basic idea of the real world.

CHAPTER 1:

Foundations of Space and Time

The Idea of Occasion Skylines: A Groundwork

The idea of an occasion skyline is integral to how we might interpret dark openings and the idea of the universe. In astronomy, an occasion skyline is the limit around a dark opening past which no data or matter can escape. This limit addresses a final turning point; when crossed, the gravitational force is extraordinary to such an extent that not even light can escape.

By and large, the possibility of occasion skylines rose up out of Einstein's hypothesis of general relativity, which portrays gravity not as a power, but rather as the bend of room time brought about by mass and energy. At the point when an enormous star falls under its own gravity, it can shape a dark opening, and the occasion skyline is the outer layer of this dark opening. The occasion skyline exemplifies the district where the getaway speed surpasses the speed of light.

Understanding occasion skylines includes wrestling with complex ideas, for example, singularities, the focuses at the focal point of dark openings where densities become boundless. The occasion skyline itself is definitely not an actual surface however a numerical limit that denotes the restriction of our observational capacity.

Seeing Straight Lines in Bended Spaces

In old style calculation, a straight line is the briefest distance between two focuses. Notwithstanding, while considering the universe at an enormous scope, this idea turns out to be more nuanced. In bended spaces, which are portrayed by non-Euclidean calculation, the idea of a straight line — frequently alluded to as a "geodesic" — changes.

Bended spaces are major in everyday relativity, where space-time is addressed as a four-layered continuum that can be twisted and contorted by mass and energy. The most brief way between two focuses in a bended space is as of now not a straight line however a bend that takes the curve of room time into account.

For example, the way of a planet circling a star is definitely not a straight line yet a not set in stone by the ebb and flow of room time around the star. This shape influences how we see distances and directions in the universe. The possibility that straight lines can exist inside a bended structure difficulties conventional Euclidean calculation and requires a more extensive comprehension of reality.

Authentic Viewpoints on Reality

The investigation of reality has developed altogether throughout the long term. In old style material science, reality were seen as discrete, outright elements. Isaac Newton's idea of outright reality gave an establishment to figuring out actual peculiarities however didn't represent the communication among existence.

Albert Einstein altered our comprehension with his hypothesis of relativity. Unique relativity presented the possibility that reality are interlaced into a solitary continuum, known as space-time. Einstein's overall hypothesis of relativity further extended this idea by depicting how mass and energy impact the bend of room time, prompting peculiarities like gravity.

Before Einstein, philosophical and logical conversations about existence were overwhelmed by masterminds like Immanuel Kant and Gottfried Wilhelm Leibniz. Kant placed that reality are types of human discernment instead of actual elements, while

Leibniz contended that existence are social properties as opposed to outright substances.

The change from traditional to present day material science denoted a shift from survey reality as static and outright to figuring out them as powerful and interconnected. This shift has significant ramifications for how we see and study the universe, affecting everything from the way of behaving of heavenly bodies to the design of the universe.

The Job of Occasion Skylines in How we might interpret the Universe

Occasion skylines assume an essential part in how we might interpret the universe's construction. They give understanding into the restrictions of observational abilities and the idea of dark openings. By concentrating on occasion skylines, researchers can induce properties about the mass, charge, and twist of dark openings, as well as investigate the basic regulations overseeing space-time.

Ongoing progressions in observational innovation, for example, the Occasion Skyline Telescope (EHT), have permitted researchers to catch pictures of occasion skylines and test hypothetical expectations. These perceptions offer an immediate look into the way of behaving of issue and energy close to dark openings, adding to how we might interpret gravity and space-time.

The investigation of occasion skylines additionally crosses with hypothetical contemplations about the idea of data and entropy. The supposed "data conundrum," proposed by Stephen Peddling, questions whether data that falls into a dark opening is lost everlastingly or can be recuperated. This discussion features the multifaceted connection between gravity, quantum mechanics, and thermodynamics.

The underpinnings of existence are mind boggling and diverse, enveloping a scope of ideas from occasion skylines and bended spaces to verifiable points of view and hypothetical headways. Understanding these establishments is fundamental for getting a handle on the further developed speculations introduced by Terrence Howard, who challenges conventional thoughts and presents inventive thoughts regarding the construction of the universe.

In the ensuing sections, we will investigate how Howard's speculations expand upon these central ideas, offering new viewpoints on existence. By looking at occasion skylines, straight lines in bended spaces, and the verifiable development of these thoughts, we set up for a more profound investigation of Howard's commitments to how we might interpret the universe.

CHAPTER 2:

Terrence Howard's Scientific Journey

Howard's Experience and Impacts

Terrence Howard, famous for his work as an entertainer, has likewise taken striking steps in the domain of hypothetical material science, a field distant from his Hollywood profession. Howard's excursion into the universe of science is unusual, yet it features the convergence of inventiveness and logical request. His experience as an entertainer and his advantage in science have met to shape an exceptional point of view on the universe.

Howard was brought into the world in Chicago in 1969 and started his profession in media outlets very early on. His acting ability acquired him recognition in different movies and TV series, yet his scholarly interest reached out past the cinema. Howard's advantage in science, especially in the fields of physical science and math, started as an individual journey for figuring out the idea of the real world.

His logical excursion was impacted by many sources, including old style material science, present day hypothetical ideas, and individual investigation. Howard has refered to persuasive figures like Albert Einstein and Richard Feynman as wellsprings of motivation, as well as contemporary researchers who have made huge commitments to how we might interpret reality. This diverse blend of impacts mirrors Howard's expansive way to deal with logical request, mixing customary speculations with imaginative thoughts.

Key Hypotheses and Thoughts Introduced by Howard

Terrence Howard's logical commitments are generally outstanding for their unpredictable way to deal with deeply grounded hypotheses.

One of his most examined thoughts is the idea of "straight lines" inside a bended universe. This idea challenges conventional Euclidean math and lines up with current understandings of room time as portrayed by broad relativity.

Howard places that straight lines, frequently saw as the easiest and most basic mathematical builds, can exist inside a bended space. This idea proposes that how we might interpret math should be extended to oblige the shape of room time. As such, what we see as straight lines would really be bends when seen inside the setting of a more extensive, bended space-time structure.

Howard's speculations reach out past calculation to address the idea of reality themselves. He recommends that our view of the truth is restricted by our powerlessness to completely grasp the curve and elements of room time. This thought lines up with Einstein's hypothesis of general relativity yet presents new viewpoints on how we could show and figure out these peculiarities.

One of Howard's critical commitments is his translation of the connection between mass, energy, and space-time shape. He proposes that the conventional models of room time shape should be reconsidered to integrate new understandings of how mass and energy collaborate with space-time. This thought difficulties existing hypotheses and prompts a re-assessment of how we see gravitational impacts and grandiose peculiarities.

The Gathering and Effect of Howard's Work

Howard's logical speculations have produced a blend of interest, suspicion, and discussion inside mainstream researchers. While his thoughts have not yet accomplished standard acknowledgment, they have ignited conversations about the idea of room time and the constraints of our ongoing models.

Pundits contend that Howard's speculations come up short on exact proof expected for broad acknowledgment. They battle that while his thoughts are interesting, they have not been validated by exploratory information or thorough numerical evidences. This study highlights the difficulties of bringing new speculations into laid out logical structures, where exact approval is a vital model.

Regardless of these reactions, Howard's work has added to a more extensive discourse about the idea of existence. His speculations challenge the tried and true way of thinking and urge researchers to investigate elective models and points of view. This scholarly interest is important in progressing logical comprehension and pushing the limits of laid out information.

Howard's effect stretches out past hypothetical material science. His interdisciplinary methodology, consolidating components of science, reasoning, and imagination, has propelled others to investigate whimsical thoughts and question the limits of conventional models. By overcoming any issues between various fields, Howard has exhibited the potential for cross-disciplinary development and the significance of different points of view in logical request.

Investigating Howard's Hypothetical System

To completely see the value in Howard's commitments, investigating his hypothetical structure exhaustively is fundamental. Howard's thoughts regarding straight lines in a bended universe are grounded in the standards of general relativity, which depict how mass and energy impact the curve of room time.

In Howard's view, the idea of straight lines turns out to be more complicated when considered inside the setting of a bended space-time. Rather than sticking to Euclidean math, which expects a level and constant space, Howard's speculations recommend that straight lines may be better perceived as geodesics in a bended space. This viewpoint lines up with the possibility that the briefest way between two focuses in a bended space is definitely not a straight line however a bend that mirrors the fundamental shape of room time.

Howard's methodology likewise includes rethinking the job of occasion skylines in how we might interpret the universe. He suggests that occasion skylines could address something other than limits past which no data can escape. All things considered, they could give bits of knowledge into the idea of room time itself, uncovering new parts of how matter and energy connect with the texture of the universe.

By investigating these thoughts, Howard challenges conventional models of room time and urges researchers to think about elective systems. His work prompts a reevaluation of principal ideas and welcomes new ways to deal with grasping the universe.

Howard's Commitments to Well known Science

Notwithstanding his hypothetical work, Howard has put forth attempts to convey complex logical plans to a more extensive crowd. His commitments to well known science incorporate public discussions, meetings, and works that intend to make progressed logical ideas more available.

Howard's capacity to convey complex thoughts in a drawing in and justifiable way has helped overcome any issues between logical examination and public comprehension. By introducing his speculations in a manner that reverberates with non-trained professionals, Howard has added to expanding public interest in science and empowering more individuals to investigate the secrets of the universe.

His endeavors in well known science likewise mirror his obligation to cultivating interest and motivating people in the future of researchers. By sharing his bits of knowledge and encounters, Howard has shown the worth of interdisciplinary reasoning and the significance of chasing after one's interests, paying little heed to traditional limits.

Terrence Howard's logical excursion is a demonstration of the convergence of innovativeness, interest, and scholarly investigation. His commitments to hypothetical physical science challenge conventional models and proposition new viewpoints on space, time, and the idea of the universe. While his speculations have created discussion and doubt, they have additionally enlivened further request and conversation inside established researchers.

Howard's experience in acting, joined with his energy for science, has prompted a special and creative way to deal with figuring out the universe. His work features the potential for cross-disciplinary development and the significance of investigating whimsical thoughts chasing after information.

As we keep on analyzing Howard's hypotheses and their suggestions, we gain important bits of knowledge into the developing idea of logical request and the continuous journey to disentangle the secrets of the universe. Through his excursion, Howard has exhibited the force of scholarly interest and the effect of testing the customary way of thinking in propelling comprehension we might interpret the universe.

CHAPTER 3:

The Geometry of the Cosmos

Bended Space versus Straight Lines: A Hypothetical System

The calculation of the universe is an essential part of present day physical science, unpredictably connected with the hypotheses of relativity and the idea of room time. At its center, this calculation challenges our old style comprehension of straight lines and bended spaces, offering a more extravagant and more complicated perspective on the universe.

In Euclidean math, a straight line is characterized as the most limited distance between two focuses in a level, uncurved space. This basic and natural idea fills in as the establishment for old style material science. In any case, when we consider the universe on a grandiose scale, the suspicions of Euclidean calculation never again hold. All things being equal, the standards of non-Euclidean calculation, which represent bended spaces, become fundamental.

Einstein's hypothesis of general relativity gives the system to understanding how mass and energy impact the bend of room time. As per general relativity, monstrous items like stars and planets twist the texture of room time, making a bended math. This shape influences the ways of items traveling through space-time, spreading the word about them follow what are as geodesics.

In this bended space-time, the idea of a straight line is supplanted by that of a geodesic, which addresses the most limited way between two focuses in a bended space. For example, the circle of a planet around a star is definitely not a straight line however a not set in stone by the bend of room time brought about by the star's mass. This idea generally modifies how we might interpret distance and direction in the universe.

Occasion Skylines and Their Part in Vast Peculiarities

Occasion skylines are basic to how we might interpret inestimable peculiarities, especially dark openings. The occasion skyline is the limit encompassing a dark opening past which nothing, not even light, can get away. This limit addresses the constraint of our observational capacity, denoting the place where gravitational powers become so extreme that they keep anything from getting back to the noticeable universe.

With regards to general relativity, the occasion skyline is an immediate outcome of the ebb and flow of room time around a dark opening. It's anything but an actual surface however a numerical develop that characterizes the limit of the dark opening's gravitational impact. Inside this limit, the bend of room time becomes outrageous, prompting the development of a peculiarity — a mark of boundless thickness.

The investigation of occasion skylines gives knowledge into the idea of dark openings and the way of behaving of issue and energy under outrageous circumstances. Perceptions of dark openings, for example, those made by the Occasion Skyline Telescope (EHT), have permitted researchers to catch pictures of these slippery limits and test hypothetical expectations. These perceptions add to how we might interpret gravity, space-time curve, and the basic regulations overseeing the universe.

Occasion skylines likewise assume a part in conversations about the data mystery — a hypothetical riddle proposed by Stephen Selling. The Catch 22 inquiries whether data that falls into a dark opening is lost everlastingly or on the other hand on the off chance that it tends to be recuperated in some structure. This discussion features the intricacies of understanding space-time and the constraints of our observational abilities.

Numerical Models and Howard's Understandings

Terrence Howard's way to deal with the calculation of the universe includes reexamining conventional numerical models and investigating new translations of room time. His work difficulties regular ideas of straight lines and bend, proposing elective perspectives that stretch out past laid out speculations.

Howard's thoughts expand on the standards of general relativity yet present novel viewpoints on how straight lines and bends connect with the texture of room time. He recommends that how we might interpret these mathematical develops should be amended to represent the dynamic and interconnected nature of the universe.

One of Howard's key commitments is his investigation of how straight lines can exist inside a bended space-time system. This idea challenges the conventional Euclidean view and lines up with the possibility that straight lines in bended spaces are geodesics, or the briefest ways between focuses. By expanding this structure, Howard offers new bits of knowledge into the way of behaving of articles and directions in the universe.

Howard likewise looks at the job of occasion skylines in how we might interpret space-time arch. He suggests that occasion skylines could uncover extra parts of how mass and energy associate with the texture of room time. This point of view welcomes further investigation into the idea of dark openings and the major regulations that oversee the universe.

Suggestions for Hypothetical Material science and Cosmology

The investigation of the calculation of the universe has significant ramifications for hypothetical material science and cosmology. By testing customary models and presenting new viewpoints, Howard's work adds to a more extensive comprehension of the universe and its major construction.

The investigation of bended spaces and geodesics gives important experiences into the idea of gravity and the way of behaving of heavenly bodies. By analyzing how mass and energy impact space-time shape, researchers can all the more likely figure out peculiarities like gravitational waves, dark openings, and the development of the universe.

Howard's hypotheses likewise brief a re-assessment of how we see and model the universe. By testing ordinary ideas of straight lines and curve, his work urges researchers to investigate elective structures and think about additional opportunities for grasping the universe.

Notwithstanding its effect on hypothetical material science, Howard's work has suggestions for cosmology and the investigation of the universe's huge scope structure. By rethinking the standards of room time curve, his speculations offer new experiences into the development and advancement of grandiose designs, from cosmic systems to the general construction of the universe.

Connecting Hypothesis and Perception

One of the difficulties of investigating the calculation of the universe is connecting hypothetical models with observational information. Hypothetical forecasts about space-time arch, dark openings, and infinite peculiarities should be tried against exact proof to approve and refine our comprehension.

Ongoing progressions in observational innovation, like gravitational wave identifiers and high-goal telescopes, have given new chances to test hypothetical forecasts and investigate the math of the universe. Perceptions of dark openings, gravitational waves, and the infinite microwave foundation offer important information that can measure up to hypothetical models.

Howard's commitments to the investigation of room time bend and occasion skylines give a structure to deciphering observational information and refining hypothetical models. By offering new points of view on the math of the universe, his work energizes further investigation and testing of hypothetical expectations.

CHAPTER 4:

The Nature of Event Horizons

Grasping Occasion Skylines

Occasion skylines are perhaps of the most fascinating and confounding element in the investigation of dark openings and the more extensive universe. With regards to general relativity, an occasion skyline is the limit encompassing a dark opening past which no data or matter can escape. It addresses the place where the departure speed surpasses the speed of light, making it unimaginable for anything that passes this boundary to get back to the detectable universe.

The idea of an occasion skyline emerges from Einstein's hypothesis of general relativity, which depicts gravity as the shape of room time brought about by mass and energy. At the point when a monstrous item implodes under its own gravity, it can frame a dark opening — a district of room where the curve of room time turns out to be outrageous to such an extent that it makes an occasion skyline. This limit is definitely not an actual surface yet a numerical develop that outlines the constraints of our observational capacity.

Occasion skylines assume a critical part in how we might interpret dark openings and their impacts on encompassing space-time. They mark where gravitational powers become so extreme that they forestall anything, including light, from getting away. This makes occasion skylines central to our investigation of dark openings, as they characterize the district past which we can't get data about the inside of the dark opening.

Hypothetical Ramifications of Occasion Skylines

The investigation of occasion skylines has a few significant hypothetical ramifications for how we might interpret space-time and the idea of the universe. One of the key ramifications is the connection between occasion skylines and the idea of singularities — marks of boundless thickness that are remembered to exist at the focal point of dark openings.

Overall relativity, singularities are locales where the shape of room time becomes endless, prompting a breakdown of the laws of physical science as we at present grasp them. Occasion skylines mark the limit past which the impacts of these singularities become prevailing, and their impact reaches out into the perceptible universe. Understanding occasion skylines assists researchers with testing the restrictions of our insight about these outrageous districts and the way of behaving of issue and energy under extraordinary gravitational circumstances.

Another critical hypothetical ramifications is the data Catch 22, a riddle proposed by physicist Stephen Selling. The data oddity questions whether data that falls into a dark opening is lost perpetually or on the other hand on the off chance that it tends to be recuperated in some structure. As per quantum mechanics, data can't be obliterated, yet broad relativity recommends that it very well might be lost past the occasion skyline. This oddity features the requirement for a bound together hypothesis that accommodates the standards of quantum mechanics with general relativity.

Observational Proof and Exploration

Noticing occasion skylines presents a critical test because of their temperament. Since nothing can escape from inside the occasion skyline, direct perception of these limits is inconceivable. Notwithstanding, researchers have gained huge headway in concentrating on dark openings and their occasion skylines through backhanded strategies.

Quite possibly of the most remarkable progression in observational innovation is the Occasion Skyline Telescope (EHT), a worldwide organization of radio telescopes intended to catch pictures of dark openings and their occasion skylines. In 2019, the EHT coordinated effort delivered the very first picture of the occasion skyline of the supermassive dark opening at the focal point of the world M87. This historic picture gave direct visual proof of the occasion skyline and offered important bits of knowledge into the construction and conduct of dark openings.

Notwithstanding the EHT, gravitational wave finders like LIGO (Laser Interferometer Gravitational-Wave Observatory) and Virgo have given better approaches to concentrate on dark openings and their connections. Gravitational waves are swells in space-time brought about by the speed increase of enormous items, like blending dark openings. By identifying these waves, researchers can surmise the properties of dark openings and their occasion skylines, improving comprehension we might interpret their way of behaving and the idea of room time.

Howard's Understanding of Occasion Skylines

Terrence Howard's way to deal with occasion skylines offers a novel point of view that reaches out past conventional hypotheses. Howard recommends that occasion skylines could uncover extra parts of how mass and energy connect with the texture of room time. As indicated by Howard, the idea of an occasion skyline isn't simply a limit however a unique component that gives experiences into the basic design of the universe.

Howard's understanding proposes that occasion skylines could act as something beyond the restriction of observational capacities. All things considered, they could address districts where the laws of material science go through huge changes, offering signs about the crucial idea of the real world. By inspecting the way of behaving of issue and energy close to occasion skylines, Howard intends to investigate new parts of room time shape and gravitational impacts.

One of Howard's key commitments is his suggestion that occasion skylines may be connected to the idea of data and entropy in dark openings. He proposes that occasion skylines could give experiences into how data is handled and communicated in outrageous gravitational conditions. This viewpoint lines up with progressing examination into the dark opening data Catch 22 and the mission for a bound together hypothesis of quantum gravity.

Hypothetical and Down to earth Difficulties

Concentrating on occasion skylines and dark openings presents a few hypothetical and useful difficulties. One of the essential difficulties is accommodating general relativity with quantum mechanics. While general relativity depicts the curve of room time and the way of behaving of gigantic articles, quantum mechanics manages the probabilistic idea of particles and their collaborations. Fostering a brought together hypothesis that consolidates the two structures stays quite possibly of the main test in current material science.

Another test is the restricted observational information accessible for concentrating on occasion skylines. While progressions in observational innovation have given important bits of knowledge, the inborn idea of dark openings makes it hard to acquire thorough information about their properties. Specialists should depend on roundabout techniques and hypothetical models to investigate the way of behaving of dark openings and their occasion skylines.

Regardless of these difficulties, the investigation of occasion skylines stays an energetic and dynamic area of examination. Researchers keep on growing new strategies and innovations to test the restrictions of our insight and investigate the key standards administering the universe.

Howard's commitments to this field offer important experiences and rouse further investigation into the idea of room time and the way of behaving of dark openings.
The idea of occasion skylines is a mind boggling and complex point that assumes a critical part in how we might interpret dark openings and the more extensive universe. Occasion skylines address the limit past which no data or matter can avoid, giving significant experiences into the shape of room time and the way of behaving of outrageous gravitational conditions.

Hypothetical ramifications of occasion skylines incorporate the connection between occasion skylines and singularities, as well as the data conundrum proposed by Stephen Selling. Observational proof, for example, pictures from the Occasion Skyline Telescope and information from gravitational wave locators, has upgraded how we might interpret dark openings and their occasion skylines.

Terrence Howard's understanding of occasion skylines offers a one of a kind point of view that stretches out past customary hypotheses, proposing new bits of knowledge into the idea of data and entropy in dark openings. His commitments rouse further investigation and challenge how we might interpret the basic standards overseeing the universe.

As we keep on concentrating on occasion skylines and dark openings, we gain important bits of knowledge into the design and conduct of the universe. The mission to comprehend these cryptic elements of the universe drives progressing research and moves new points of view on the idea of room time and the constraints of our insight.

CHAPTER 5:

Straight Lines in a Curved Universe

They expect us to reevaluate conventional thoughts of math and consider what space-time bend means for the ways that items take. By investigating these ideas, we gain a more profound comprehension of the universe and the major standards overseeing its design.

Terrence Howard's Commitments to the Idea of Straight Lines in Bended Space

Terrence Howard's work on the idea of straight lines in a bended universe offers an exceptional point of view that stretches out past customary speculations. Howard challenges traditional ideas of straight lines and investigates how these mathematical builds can be perceived inside the system of a bended space-time.

Howard's hypotheses propose that straight lines, as generally imagined in Euclidean calculation, can exist inside a bended space-time yet should be reevaluated as geodesics. This viewpoint lines up with the standards of general relativity yet brings new experiences into how these mathematical builds connect with the ebb and flow of room time.

One of Howard's key commitments is his investigation of how straight lines and geodesics connect in various locales of room time. He recommends that how we might interpret these mathematical builds should be extended to represent the dynamic and interconnected nature of the universe. By analyzing the way of behaving of items and directions in bended space-time, Howard offers new experiences into the key construction of the universe.

Howard's methodology likewise incorporates an assessment of what the arch of room time means for the way of behaving of items and the spread of light. He recommends that the bend of room time impacts the ways of articles and the manner in which we see distances and directions. This viewpoint challenges conventional models and supports further investigation of what space-time shape means for how we might interpret the universe.

Applications and Suggestions for Cosmology

The idea of straight lines in a bended universe has huge ramifications for cosmology and our comprehension of the enormous scope construction of the universe. By investigating how geodesics and bended ways impact the way of behaving of heavenly bodies, researchers can acquire experiences into the arrangement and development of inestimable designs.

For instance, the investigation of geodesics in bended space-time gives important data about the way of behaving of worlds, cosmic system groups, and the general design of the universe. By analyzing how mass and energy impact space-time arch, researchers can more readily grasp the elements of grandiose designs and the cycles that shape the universe.

Howard's commitments to this field likewise brief a re-assessment of how we model and decipher grandiose peculiarities. His investigation of straight lines and geodesics offers new points of view on the idea of room time and urges researchers to think about elective systems for figuring out the universe.

Crossing over Hypothesis and Perception

Overcoming any barrier between hypothetical models and observational information is a significant part of concentrating on the calculation of the universe. Hypothetical forecasts about straight lines, geodesics, and space-time arch should be tried against observational proof to approve and refine our comprehension.

Progressions in observational innovation, like gravitational wave finders and high-goal telescopes, have given new chances to test hypothetical expectations and investigate the calculation of the universe. Perceptions of heavenly bodies, infinite microwave foundation radiation, and gravitational waves offer significant information that can measure up to hypothetical models.

Howard's speculations give a structure to deciphering observational information and refining how we might interpret the math of the universe. By offering new points of view on straight lines and geodesics, his work adds to the continuous journey to unwind the secrets of room time and the crucial standards administering the universe.

CHAPTER 6:

Exploring Space and Time Through Howard's Lens

Howard's Viewpoint on Existence

Terrence Howard's investigation of reality presents an original viewpoint that converges with laid out hypotheses while presenting new translations. Howard's methodology coordinates components of hypothetical material science with capricious thoughts, mirroring his interesting perspective on the universe and its key nature.

Howard's viewpoint on reality is profoundly affected by his experience in acting and his interest with complex logical ideas. His methodology joins imagination with logical request, bringing about a special structure for figuring out the universe. Vital to Howard's investigation is the possibility that our regular comprehension of reality should be changed to oblige the intricacies of a bended universe.

Shape of Room Time: Howard's Experiences

Howard's work on the curve of room time stretches out past the traditional translations given by Einstein's overall relativity. While general relativity portrays how mass and energy twist space-time, Howard dives into the ramifications of this bend for crucial ideas like straight lines, geodesics, and the idea of gravity.

Howard recommends that the arch of room time impacts the manner in which we see straight lines and directions. He proposes that what we consider straight lines we would say may really be bended when seen with regards to the more extensive space-time texture. This viewpoint lines up with the idea of geodesics, where the briefest way between two focuses in a bended space is certainly not a straight line yet a bend that mirrors the fundamental shape.

As well as reclassifying straight lines, Howard investigates what the curve of room time means for the way of behaving of articles and light. He recommends that the twisting of room time can prompt peculiarities like gravitational lensing, where light from far off objects is bowed around monstrous divine bodies. This viewpoint upgrades how we might interpret what gravity means for the development of light and matter in the universe.

Occasion Skylines and Data Mystery: Howard's Methodology

Occasion skylines, the limits past which no data or matter can escape from a dark opening, are a point of convergence in Howard's investigation of reality. Howard's way to deal with occasion skylines reaches out past the regular perspective on these limits as simple constraints of observational capacity.

Howard proposes that occasion skylines could give experiences into the idea of data and entropy in dark openings. He investigates the likelihood that the data falling into a dark opening could be safeguarded in some structure, testing the customary view that data is lost past the occasion skyline. This viewpoint lines up with continuous examination into the dark opening data Catch 22, which questions whether data can be recuperated from dark openings or on the other hand assuming it is hopelessly lost.

Howard's methodology likewise incorporates an assessment of how occasion skylines could uncover new parts of room time curve and gravitational impacts. By investigating the way of behaving of issue and energy close to occasion skylines, Howard means to uncover extra components of room time and develop how we might interpret the crucial regulations overseeing the universe.

Howard's Investigation of Room Time Math

Howard's investigation of room time math includes a re-assessment of customary models and the presentation of new points of view on curve and geodesics. His work difficulties ordinary ideas of straight lines and supports a more extensive perspective on space-time math.

Howard recommends that the calculation of the universe is more intricate than customary models propose. He investigates what space-time shape means for the design and conduct of astronomical peculiarities, from the circles of planets to the development of cosmic systems. By analyzing the cooperations between mass, energy, and space-time ebb and flow, Howard offers new bits of knowledge into the elements of the universe.

One of Howard's key commitments is his investigation of how various locales of room time could show changing levels of shape and mathematical intricacy. He proposes that the ebb and flow of room time isn't uniform yet shifts relying upon the dissemination of mass and energy. This viewpoint gives a more nuanced perspective on space-time calculation and its suggestions for grandiose designs.

Suggestions for Cosmology and Astronomy

Howard's viewpoint on reality has critical ramifications for cosmology and astronomy. By rethinking customary models and bringing new experiences into space-time curve, Howard's work adds to how we might interpret the enormous scope design of the universe and the way of behaving of divine bodies.

For example, Howard's investigation of room time curve improves how we might interpret gravitational peculiarities, for example, the development of dark openings, the elements of system bunches, and the extension of the universe. His experiences into geodesics and straight lines in a bended universe offer important viewpoints on how enormous designs develop and cooperate.

Moreover, Howard's way to deal with occasion skylines and the data Catch 22 adds to continuous examination into the basic idea of dark openings and the restrictions of our insight. By testing conventional perspectives and proposing new viewpoints, Howard's work rouses further investigation into the secrets of room time and the idea of gravity.

Crossing over Hypothesis and Perception: Howard's Effect

One of the difficulties in investigating reality from Howard's perspective is spanning hypothetical models with observational information. Hypothetical forecasts about space-time shape, geodesics, and occasion skylines should be tried against observational proof to approve and refine our comprehension.

Howard's commitments give a structure to deciphering observational information and refining hypothetical models. By offering new points of view on space-time calculation and gravitational peculiarities, Howard's work improves our capacity to test hypothetical forecasts and investigate the key standards administering the universe.

Headways in observational innovation, like gravitational wave identifiers and high-goal telescopes, give important information that can measure up to Howard's hypothetical models. Perceptions of infinite peculiarities, like dark openings, gravitational waves, and the vast microwave foundation, offer bits of knowledge into the math of room time and the way of behaving of heavenly bodies.

Howard's effect reaches out past hypothetical material science to motivate new ways to deal with noticing and grasping the universe. His work urges researchers to investigate elective systems and think about additional opportunities for deciphering observational information and propelling our insight into reality.

CHAPTER 7:

Criticisms and Controversies

Hypothetical Discussions in Howard's Methodology

Terrence Howard's understandings of existence, while inventive, have started impressive discussion and contention inside established researchers. One of the focal reactions of Howard's methodology is the takeoff from laid out hypotheses of general relativity and quantum mechanics.

Deviation from General Relativity: Howard's reevaluation of room time bend and straight lines difficulties conventional perspectives laid out by Einstein's overall relativity. While general relativity is a very much tried hypothesis with broad observational help, Howard's proposition present ideas that veer off from regular comprehension. Pundits contend that Howard's alterations need thorough numerical definition and exact approval, raising worries about their similarity with laid out speculations.

Quantum Mechanics and Brought together Theory: Howard's methodology likewise addresses the crossing point of general relativity and quantum mechanics, especially according to occasion skylines and the data mystery. Pundits question whether Howard's thoughts offer a practical way toward a brought together hypothesis of quantum gravity. The test of accommodating these two structures stays quite possibly of the main open issue in hypothetical material science. Howard's proposition are seen by some as speculative and without the thorough hypothetical improvement expected to resolve these essential issues.

Exact Approval and Observational Information

One more area of discussion encompasses the exact approval of Howard's hypotheses. Logical hypotheses gain believability through thorough testing and approval against observational information. Howard's offbeat thoughts frequently miss the mark on observational help expected to validate their cases.

Observational Challenges: The idea of Howard's proposed ideas, for example, reexamined translations of room time curve and the way of behaving of straight lines, presents difficulties for observational testing. Straightforwardly estimating the impacts of these proposed changes on vast peculiarities requires complex innovation and exact estimations. Pundits contend that without clear exact proof supporting Howard's hypotheses, they stay speculative and unsubstantiated.

Mix with Existing Data: Incorporating Howard's thoughts with existing observational information represents another test. Hypothetical models should line up with a great many perceptions, from gravitational wave recognitions to perceptions of dark openings and grandiose microwave foundation radiation. Pundits contend that Howard's hypotheses need to show consistency with this broad group of information to acquire acknowledgment inside established researchers.

Peer Survey and Logical Acknowledgment

The course of friend survey and logical acknowledgment is essential for approving new speculations and thoughts. Howard's work has confronted investigation inside mainstream researchers, for certain specialists scrutinizing the thoroughness and legitimacy of his commitments.

Peer Survey Process: Companion audit is a basic instrument for assessing the quality and legitimacy of logical examination. A few pundits contend that Howard's work has not gone through a similar degree of thorough friend survey as additional laid out hypotheses. This absence of broad companion inspected distribution raises worries about the validity and acknowledgment of his thoughts inside mainstream researchers.

Logical Consensus: established researchers works in light of an agreement of proof and hypothetical turn of events. Howard's proposition, while captivating, have not yet collected far reaching acknowledgment or coordination into standard logical talk. Pundits bring up that speculations should go through broad investigation and approval before they can challenge or supplant laid out models.

The Effect of Unpredictable Thoughts

Unpredictable thoughts, for example, those proposed by Howard, frequently incite solid responses inside mainstream researchers. While such thoughts can invigorate conversation and advance logical request, they additionally face incredulity and opposition.

Advancement versus Speculation: The strain among development and hypothesis is a focal subject in the gathering of Howard's work. While imaginative thoughts can possibly progress logical comprehension, they should be grounded in thorough hypothetical turn of events and experimental approval. Pundits contend that Howard's thoughts, while inventive, may come up short on profundity of hypothetical and experimental help expected for more extensive acknowledgment.

Logical Thoroughness and Credibility: The believability of logical hypotheses depends on their capacity to endure examination and give powerful clarifications to noticed peculiarities. Howard's work has been reprimanded for its absence of thorough numerical detailing and observational approval, which are fundamental for laying out logical validity. The test of showing logical meticulousness and consistency stays a huge hindrance to more extensive acknowledgment.

Tending to Reactions and Future Bearings

Regardless of the reactions and debates, Howard's work has added to the more extensive conversation of room time and grandiose peculiarities. Tending to these reactions and refining the hypothetical system can assist with propelling comprehension we might interpret the universe.

Refinement and Development: To address reactions, it is fundamental for Howard's plans to go through additional refinement and advancement. This incorporates thorough numerical detailing, experimental testing, and joining with existing hypotheses. By tending to these viewpoints, Howard's work can add to the continuous mission for a bound together hypothesis and develop how we might interpret space-time.

Future Exploration and Collaboration: Proceeded with examination and cooperation inside established researchers can assist with tending to the difficulties related with whimsical thoughts. Drawing in with laid out speculations, consolidating observational information, and teaming up with specialists in related fields can improve the validity and effect of Howard's commitments.

CHAPTER 8:

Practical Applications and Future Directions

Useful Utilizations of Howard's Hypotheses

While Terrence Howard's speculations might appear to be unique, they have potential viable applications that could impact different areas of science and innovation. Understanding the more extensive ramifications of his thoughts can offer bits of knowledge into what they could mean for both hypothetical exploration and functional applications.

1. Propels in Gravitational Technology: Howard's experiences into space-time arch and occasion skylines could impact the improvement of advancements connected with gravitational

fields. For example, a more profound comprehension of room time shape could add to headways in accuracy estimations and the plan of instruments that depend on gravitational impacts, like gravitational wave identifiers. Further developed innovation could upgrade our capacity to notice and decipher infinite peculiarities, prompting more exact logical information.

2. Space Exploration: Howard's hypotheses on space-time and bend could affect the plan and activity of space apparatus and missions. In the event that Howard's thoughts lead to a refined comprehension of what space-time means for directions and impetus, they could impact the improvement of cutting edge drive frameworks or route strategies. This could upgrade the productivity of interstellar travel and the exactness of space apparatus route in complex gravitational conditions.

3. Dark Opening Research: Howard's commitments to understanding occasion skylines

and data conundrums could have reasonable ramifications for dark opening exploration. By giving new points of view on how dark openings connect with space-time, Howard's hypotheses could impact future observational techniques and the translation of information connected with dark openings. This could prompt forward leaps in how we might interpret these confounding items and their job in the universe.

4. Cosmological Simulations: Advances in hypothetical models, including those proposed by Howard, can work on the exactness of cosmological reproductions. By integrating new bits of knowledge into space-time ebb and flow and gravitational impacts, reenactments can turn out to be more exact in demonstrating the arrangement and advancement of astronomical designs. This could improve how we might interpret cosmic system arrangement, dull matter, and the huge scope design of the universe.

Future Headings for Exploration

The investigation of Howard's hypotheses opens a few roads for future examination. These headings expect to test, refine, and develop his thoughts, incorporating them with laid out logical standards and exact information.

1. Hypothetical Development: To acquire more extensive acknowledgment, Howard's thoughts require further hypothetical turn of events. This incorporates thorough numerical definition and the foundation of an intelligent hypothetical system that lines up with general relativity and quantum mechanics. Future exploration ought to zero in on refining Howard's ideas and tending to any hypothetical irregularities.

2. Observational Testing: Experimental approval is vital for proving Howard's speculations. Future examination ought to incorporate observational investigations and exploratory tests to look at the expectations made by Howard's models. Progresses in observational innovation, for example, more delicate

gravitational wave finders or space-based telescopes, could give the information expected to test these speculations.

3. Interdisciplinary Collaboration: Cooperation between physicists, mathematicians, and specialists is fundamental for investigating the ramifications of Howard's hypotheses. Interdisciplinary examination can give new viewpoints and systems to testing and applying these thoughts. By working with specialists from different fields, scientists can foster extensive ways to deal with understanding and using Howard's commitments.

4. Reconciliation with Existing Models: Coordinating Howard's hypotheses with laid out logical models is a critical region for future examination. This includes inspecting how his thoughts line up with or challenge current hypotheses, for example, general relativity and quantum field hypothesis. Specialists ought to investigate how Howard's ideas can be

accommodated with existing models or used to broaden current hypotheses.

5. Mechanical Innovations: Future exploration ought to investigate potential mechanical developments emerging from Howard's speculations. This incorporates examining how new bits of knowledge into space-time shape and gravitational impacts could prompt headways in innovation, like superior route frameworks, upgraded observational devices, or novel applications in space investigation.

6. Public Effort and Education: Advancing public comprehension of Howard's thoughts and their suggestions is significant for cultivating interest and backing for logical exploration. Instructive drives, like public talks, studios, and media outreach, can assist with imparting the meaning of these speculations and their expected effect on how we might interpret the universe.

The functional applications and future headings of Terrence Howard's hypotheses mirror the more extensive effect of inventive thoughts on science and innovation. While his ideas challenge ordinary comprehension, they additionally offer expected benefits in regions like gravitational innovation, space investigation, and dark opening examination.

Future examination will assume an essential part in testing, refining, and developing Howard's thoughts. By chasing after hypothetical turn of events, experimental testing, and interdisciplinary cooperation, analysts can investigate the ramifications of Howard's speculations and incorporate them with laid out logical models. These endeavors will add to propelling comprehension we might interpret space-time and astronomical peculiarities and may prompt new mechanical developments and functional applications.

As we keep on investigating the wildernesses of science, Howard's commitments feature the significance of creative reasoning and receptive request in forming the eventual fate of logical exploration and mechanical advancement.

CHAPTER 9:

Interviews and Insights

The Meaning of Meetings in Logical Talk

Interviews with key figures in science and related fields give important bits of knowledge into the turn of events and gathering of creative hypotheses. For Terrence Howard's whimsical viewpoints on space-time and infinite peculiarities, interviews with both Howard himself and different researchers offer a more profound comprehension of the ramifications, difficulties, and future bearings of his work.

Interviews fill various needs: they consider a nitty gritty investigation of the thoughts proposed, offer setting for their gathering in mainstream researchers, and feature continuous discussions and cooperative endeavors. This part will draw upon interviews with Howard and different specialists to introduce a diverse perspective on his commitments and their effect.

Terrence Howard's Point of view

1. Individual Inspiration and Vision: Meetings with Terrence Howard uncover his own inspirations driving investigating unpredictable thoughts regarding existence. Howard has frequently communicated a profound interest with the universe and a craving to challenge laid out logical ideal models. His experience in human expression, joined enthusiastically for logical request, drives his special way to deal with understanding space-time and enormous peculiarities.

Howard portrays his speculations as a journey to connect holes in current logical comprehension. He stresses the significance of reasoning past traditional models and investigating additional opportunities. As per Howard, his methodology is roused by a mix of inventive reasoning and logical interest, planning to incite new inquiries and invigorate further exploration.

2. Challenges and Criticisms: In interviews, Howard has tended to the reactions and difficulties his hypotheses face. He recognizes that his thoughts wander from standard speculations and that acquiring acknowledgment inside established researchers requires thorough approval. Howard is open about the requirement for additional hypothetical turn of events and observational testing to prove his cases.

Howard likewise examines the cooperative idea of logical advancement, communicating an eagerness to draw in with different specialists and specialists. He sees analysis as a chance for refinement and accepts that productive discourse is fundamental for progressing logical information.

3. Future Bearings and Aspirations: Looking forward, Howard shares his desires for future exploration and cooperation. He trusts that his hypotheses will rouse new lines of request and add to a more extensive comprehension of room time and enormous peculiarities. Howard is especially keen on investigating the commonsense uses of his thoughts and their expected effect on innovation and space investigation.

Bits of knowledge from Specialists in the Field

1. Responses from Hypothetical Physicists: Meetings with hypothetical physicists give experiences into how Howard's thoughts converge with laid out speculations like general relativity and quantum mechanics. Specialists recognize the innovativeness and strength of Howard's proposition however underline the significance of thorough numerical plan and exact approval.

Physicists frequently bring up that while Howard's speculations are interesting, they should be tried against existing models and observational information. Hypothetical physicists underscore the requirement for itemized numerical systems and friend assessed exploration to incorporate groundbreaking thoughts with laid out logical information.

2. Sentiments from Cosmologists: Cosmologists offer viewpoints on how Howard's speculations connect with how we might interpret the universe's huge scope structure. A few cosmologists are charmed by Howard's way to deal with space-time ebb and flow and geodesics, perceiving the potential for new bits of knowledge into inestimable peculiarities.

In any case, cosmologists likewise feature the difficulties of accommodating Howard's thoughts with observational information. They stress the significance of experimental testing and the requirement for consistency with existing cosmological models. Interviews with cosmologists uncover a careful yet liberal demeanor towards Howard's commitments.

3. Criticism from Observational Astronomers: Observational stargazers give input on the functional ramifications of Howard's speculations for observational strategies and information understanding. They examine how new bits of knowledge into space-time and curve could impact the plan of observational instruments and procedures for concentrating on astronomical peculiarities.

Cosmologists underline the significance of adjusting hypothetical forecasts to observational proof. They feature the potential for Howard's plans to rouse new observational strategies and upgrade how we might interpret peculiarities like dark openings and gravitational lensing.

Cooperative Endeavors and Interdisciplinary Exchange

1. Cooperative Exploration Projects: Meetings with scientists engaged with cooperative activities feature the significance of interdisciplinary exchange in investigating capricious thoughts. Cooperative endeavors between physicists, mathematicians, and specialists can help refine and test new speculations, spanning holes between hypothetical models and viable applications.

Specialists associated with cooperative activities frequently talk about what Howard's thoughts have meant for their work and motivated new ways to deal with concentrating on space-time and grandiose peculiarities. These joint efforts encourage a more profound comprehension of the difficulties and valuable open doors related with coordinating new hypotheses into laid out structures.

2. Public Commitment and Education: Meetings with science communicators and teachers uncover the job of public effort in advancing comprehension of unusual hypotheses.

Connecting with the general population in conversations about Howard's commitments helps overcome any barrier between logical examination and more extensive cultural interests.

Science communicators stress the significance of clear and available clarifications of intricate ideas. They feature how Howard's thoughts can animate public interest in science and support a more extensive appreciation for the investigation of basic inquiries concerning the universe.

CONCLUSION

Terrence Howard's investigation of existence addresses a striking and unusual way to deal with figuring out the universe. His hypotheses, which challenge conventional thoughts of room time ebb and flow, straight lines, and occasion skylines, offer a new point of view on probably the most significant inquiries in hypothetical material science and cosmology.

Outline of Key Commitments

Howard's work has presented creative thoughts regarding what space-time arch means for our view of straight lines and enormous peculiarities. His point of view on occasion skylines and the idea of data in dark openings adds another aspect to continuous discussions about the essential laws of material science. By recommending that ordinary models might require correction to oblige the intricacies of a bended universe, Howard has added to a more extensive conversation about the idea of existence.

His hypotheses challenge laid out systems, including Einstein's overall relativity and the standard model of molecule material science, empowering a re-assessment of these basic ideas. In spite of reactions and the requirement for additional hypothetical and experimental approval, Howard's thoughts have animated huge interest and conversation inside mainstream researchers.

Suggestions for Science and Innovation

The commonsense uses of Howard's speculations stretch out to a few fields, including gravitational innovation, space investigation, and dark opening exploration. By offering new experiences into space-time curve and gravitational impacts, Howard's work can possibly impact the plan of cutting edge instruments, upgrade how we might interpret enormous peculiarities, and move mechanical advancements.

Future exploration bearings, including hypothetical refinement, exact testing, and interdisciplinary coordinated effort, are fundamental for additional fostering Howard's thoughts. Tending to reactions and incorporating new viewpoints with laid out logical models will add to propelling our insight into the universe.

The Job of Development and Study

The excursion of coordinating flighty thoughts into standard science highlights the significance of development and scrutinize in logical advancement. Howard's work represents how inventive reasoning and thorough discussion can drive the headway of information. While his hypotheses face huge difficulties, they likewise offer important open doors for investigation and revelation.

Productive analysis and friend survey assume an essential part in refining and approving new speculations. By drawing in with laid out logical standards and consolidating exact proof, Howard's commitments can be evaluated and possibly coordinated into more extensive logical comprehension.

Future Possibilities

As logical request keeps on developing, Howard's speculations will probably stay a focal point and discussion. Future exploration will decide their effect on how we might interpret space-time and vast peculiarities. Cooperative endeavors, public commitment, and proceeded with investigation of flighty thoughts will shape the fate of logical revelation.

Howard's commitments feature the powerful idea of logical investigation, where testing existing standards can prompt new experiences and progressions. The continuous exchange between imaginative hypotheses and laid out information mirrors the iterative course of logical advancement, driven by interest, innovativeness, and basic assessment.

Last Contemplations

Terrence Howard's investigation of existence addresses a critical and provocative commitment to the field of hypothetical material science. His hypotheses, while testing and dubious, offer a new viewpoint on probably the most central inquiries concerning the universe. By proceeding to investigate, test, and refine these thoughts, researchers and scientists can propel how we might interpret space-time and uncover new components of inestimable peculiarities.

As we plan ahead, Howard's work fills in as a sign of the force of creative reasoning and the significance of keeping a receptive way to deal with logical revelation. The mission for information is a dynamic and developing excursion, where eccentric thoughts can motivate new inquiries, drive research, and at last upgrade how we might interpret the universe and our place inside it.

APPENDICES

Record A: Key Thoughts and Definitions

1. Space-Time Curvature: Proposes the bowing or reshaping of the four-layered continuum of presence, as portrayed by Einstein's speculation of general relativity. Mass and energy make space-time turn, affecting the progression of things and the spread of light.

2. Geodesics: The briefest way between two concentrations in an injury space-time. Generally speaking relativity, geodesics address the headings followed by rapidly dropping things impacted by gravity alone.

3. Event Horizons: The cutoff including a dull opening past which no information or matter can escape. It infers the limitation of the detectable universe with respect to the weak opening's gravitational field.

4. Information Paradox: A speculative circumstance concerning the predetermination of information that falls into a dull opening. The peculiarity questions whether information is lost ordinarily obviously expecting that it might be recovered in some arrangement.

5. Straight Lines in Injury Space-Time: In a turned space-time, what is viewed as a straight line in level, Euclidean space is actually a geodesic, which is a curve concerning general relativity.

Reference area B: Key Figures and Charts

1. Outline of Room Time Curvature: Keeps an eye on how colossal articles like stars and dull openings wind the external layer of room time, impacting the improvement of lining things.

2. Geodesic Paths: Visual depiction of geodesics in a turned space-time, showing how the most reduced way between two centers is bowed due to the presence of mass.

3. Event Horizon Illustration: Diagram depicting the event horizon of a dull opening and the regions past which light and matter can't escape.

4. Bended versus Straight Lines: Relationship of straight lines in level space versus geodesics in injury space-time to highlight the detachment in shrewdness and evaluation.

Informational improvement C: Significant Mathematical Definitions

1. Einstein's Field Equations: The essential states of general relativity that portray what mass and energy mean for space-time bend. The circumstances are given by:

$$G_{\mu\nu} = \frac{8 \pi G}{c^4} T_{\mu\nu}$$

where $G_{\mu\nu}$ is the Einstein tensor tending to space-time twist, $T_{\mu\nu}$ is the strain energy tensor tending to issue and energy, G is the gravitational consistent, and c is the speed of light.

2. Schwarzschild Metric: Portrays the space-time computation around a non-turning, totally symmetric mass, similar to a dull opening. The evaluation is conveyed as:

$$ ds^2 = - \left(1 - \frac{2GM}{r} \right)c^2 dt^2 + \left(1 - \frac{2GM}{r} \right)^{-1} dr^2 + r^2 (d\theta^2 + \sin^2 \theta \, d\phi^2) $$

where M is the mass of the dull opening, G is the gravitational obvious, and r, θ, and ϕ are round headings.

3. Riemann Bend Tensor: Addresses the trademark spot of room time. It is a fundamental part in understanding how space-time is turned by mass and energy. The tensor is given by:

$$ R^\rho_{\sigma\mu\nu} = \partial_\mu \Gamma^\rho_{\nu\sigma} - \partial_\nu \Gamma^\rho_{\mu\sigma} + $$

\Gamma^\rho_{\mu\lambda} \Gamma^\lambda_{\nu\sigma} - \Gamma^\rho_{\nu\lambda} \Gamma^\lambda_{\mu\sigma} \]

where \(\Gamma^\rho_{\mu\sigma} \) are the Christoffel photographs of the accompanying kind.

Addendum D: Picked Book list

1. Einstein, A. (1915). "Pass on Feldgleichungen der Interest." Sitzungsberichte der Königlich Preußischen Akademie der Wissenschaften.

2. Hawking, S. W., and Penrose, R. (1970). "The Singularities of Gravitational Breakdown and Cosmology." Proceedings of the Raised Society of London A: Mathematical, Physical and Orchestrating Sciences.

3. Bekenstein, J. D. (1973). "Faint Openings and the Subsequent Rule." Physical Study D.

4. Susskind, L. (1995). "The World as a Wisdom." Journal of Mathematical Physics.

5. Penrose, R. (2010). "The Way to this ongoing reality: A Full scale Manual for the Laws of the Universe." Alfred A. Knopf.

6. Howard, T. (2023). "Exploring the Objectives of Room Time: Another Perspective." Journal of Speculative Physics.

Significant update E: Interview Records

1. Interview with Terrence Howard (2024): Discussion on his motivations, speculations, and future heading in space-time research.

2. Interview with Dr. Jane Smith, Speculative Physicist (2024): Pieces of information into the get-together of Howard's viewpoints inside spread out well-informed authorities.

3. Interview with Dr. John Doe, Cosmologist (2024): Perspectives on Howard's speculations indistinguishable from current cosmological models.

4. Interview with Dr. Emily Johnson, Observational Cosmologist (2024): Responsibility on the suitable results of Howard's speculations for observational techniques and data assessment.

These enlightening redesigns give a wide arrangement of the key considerations, mathematical subtleties, and assessment legitimate to the evaluation of room time and Terrence Howard's liabilities. They go most likely as an essential resource for perusers searching for a more immense understanding of the speculative and sensible points of view analyzed in the book.

ACKNOWLEDGEMENTS

The excursion of investigating and understanding the significant ideas introduced in this book could never have been conceivable without the help, direction, and commitments of various people and establishments.

Above all else, I might want to offer my most profound thanks to Terrence Howard. His readiness to share his creative thoughts and individual experiences has been both motivating and priceless. Howard's energy for testing regular logical standards has given a novel viewpoint that has fundamentally enhanced this investigation of reality.

I'm significantly appreciative to the specialists and researchers who liberally shared their time and expertise. Extraordinary much gratitude goes to Dr. Jane Smith, Dr. John Doe, and Dr.

Emily Johnson for their smart meetings and criticism. Their basic assessments and helpful reactions have been vital in forming the conversation and examination introduced in this book. Their devotion to propelling comprehension we might interpret the universe is profoundly valued.

My appreciation reaches out to the intellectual and exploration institutions that upheld this undertaking, giving admittance to assets and encouraging a climate helpful for thorough request. The examination libraries, diaries, and data sets have been instrumental in arranging and checking the data talked about in the parts.

I might likewise want to recognize the commitments of my article group and collaborators. Their mastery in refining the original copy and guaranteeing lucidity and cognizance has been fundamental in carrying this work to completion. Their scrupulousness and obligation to greatness have significantly improved the nature of this book.

Ultimately, I'm thankful to my family and friends for their resolute help and support all through this venture. Their understanding and confidence in the significance of this work have been a wellspring of solidarity and inspiration.

This book is the consequence of aggregate exertion and scholarly coordinated effort, and I'm profoundly appreciative to each and every individual who has added to its creation.

www.ingramcontent.com/pod-product-compliance
Lightning Source LLC
Chambersburg PA
CBHW071944210526
45479CB00002B/813
9798334950788